Garfield™ Collectibles

Debra S. Braun

4880 Lower Valley Rd. Atglen, PA 19310 USA

GARFIELD is the property of PAWS, Inc., a licensing and creative studio located near Muncie, Indiana. GARFIELD© PAWS all rights reserved. Garfield ™ is a registered trademark of PAWS, Inc. All pictures, graphics and photos compiled herein are intended to heighten the awareness of Garfield ™ and related products. This book is in no way intended to infringe on the intellectual property rights of any party. All products, brands, and names represented are trademarks or registered trademarks of their respective companies. Information in this book was derived from the author's independent research and was not authorized, furnished, or approved by PAWS, Inc. or Jim Davis.

Copyright © 1998 by Debra S. Braun
Library of Congress Catalog Card Number:98-84277

Book design by Laurie A.Smucker
Typeset in Futura BK BT/Dombold BT

ISBN: 0-7643-0547-6
Printed in China
1 2 3 4

Dedication
This book is dedicated to
all of the Garfield fans in the world who live
for lasagna and laughter.

Acknowledgements

I would like to extend a special thanks to my parents, Mark Coté, Peter and Nancy Schiffer, Doug Congdon-Martin, Kim Campbell, and the staff at Wink One Hour Photo who were instrumental in the development of this book. I really appreciate their guidance and support.

I would also like to thank my co-workers and friends for believing in me and making me smile.

Published by Schiffer Publishing Ltd.
4880 Lower Valley Road
Atglen, PA 19310
Phone: (610) 593-1777; Fax: (610) 593-2002
E-mail: Schifferbk@aol.com
Please write for a free catalog.
This book may be purchased from the publisher.
Please include $3.95 for shipping.

In Europe, Schiffer books are distributed by
Bushwood Books
6 Marksbury Ave
Kew Gardens
Surrey TW9 4JF England
Phone: 44 (0)181 392-8585; Fax: 44 (0)181 392-9876
E-mail: Bushwd@aol.com

Please try your bookstore first.

We are interested in hearing from authors
with book ideas on related subjects.

Contents

Introduction

Garfield's creator, Jim Davis, is no stranger to cats; after all, he grew up with twenty-five of them in rural Indiana. The Garfield comic strip first appeared on June 19, 1978 and was included in approximately forty newspapers that year. Since then, it has appeared in over 2,400 newspapers worldwide in at least twelve languages. It is estimated that Garfield is read by more than a million people each day.

According to the *Garfield Trivia Book,* it takes Jim Davis approximately eight weeks to create a daily comic strip. Sunday comic strips take even longer.

In the comic strip, Garfield weighed five pounds, six ounces when he was born in an Italian restaurant. It is no wonder that he loves lasagna! It is an interesting fact that Jim Davis's son also weighed five pounds, six ounces at birth.

Garfield's life mainly revolves around the other characters in the comic strip: Jon Arbuckle, Odie, Pooky, Arlene, and Nermal.

Garfield is a bright orange cat who has an insatiable appetite and loves to take power naps.

Jon Arbuckle is a cartoonist who adopted Garfield from a pet store.

Odie is a happy-go-lucky dog that is blissfully playful. Sometimes he gets on Garfield's nerves. Do you remember when Garfield kicked Odie into next week?

Pooky is Garfield's teddy bear. He has been Garfield's confidante ever since he was discovered in the bottom drawer of a dresser. It was traumatic when Pooky's arm fell off in one of the cartoons. Garfield actually fainted! Do you remember the kidnapping ploy that Jon had to pull off so that he could sew Pooky's arm back on?

Arlene is Garfield's girlfriend. She is pink with passion and has lucious red lips.

Nermal is a gray kitten who is owned by Jon Arbuckle's parents. He is also known as "the world's cutest kitten."

Garfield's success has spawned literally thousands of collectibles throughout the years. To keep up with the marketing demand, Jim Davis founded PAWS, Inc. in 1981. It is located near Muncie, Indiana. By 1989, it grew into a 36,000 square foot facility. PAWS, Inc. currently employs 55 people.

Garfield has remained in the spotlight for many years. In 1982, Garfield's first television show, called *"Here comes Garfield,"* was aired. Then, in 1984, Garfield made his debut in the Macy's Day parade. In 1985, *"Garfield in the Rough"* was awarded an Emmy for the best animated special. An officially endorsed fan club was launched in 1997.

There are plans to open a $120 million Garfield theme park in Hendrick County, Indiana in 1999. Approximately 503 acres will house an amphitheater, rides, upcoming attractions, and shopping. Of course, a Garfield theme park would not be complete without lots of places to eat. Now that's food for thought!

To join the Garfield fan club or receive a merchandising catalog, collectors can call 1-888-274-PAWS (7297). The Garfield Web site address is: http://www.garfield.com

Helpful Hints

Searching for Garfield items at toy shows, flea markets, and garage sales is challenging and fun. You never know what you are going to find. It is important to understand that the prices for Garfield memorabilia will fluctuate according to their condition, supply, and demand. You should purchase an object because you like it, not for an investment. This book is intended to heighten your awareness of Garfield collectibles and their approximate values in excellent to mint condition on a secondary market. All the items shown in this book were produced between 1978 and 1997.

There are several things to consider before purchasing an object. For example: Is it in good condition? Are all of the pieces complete? Is it rare and hard to find? It is my recommendation to inspect each item thoroughly. Especially look for chips, hairline cracks, paint peeling, and crazing on ceramic items.

Tip 1. One way to identify a hairline crack on an object is to hold it up to a light. The hairline crack should stick out instantly even if it has been repaired.

Tip 2. To feel a chip easily, gently move your finger around the entire surface of an object.

Tip 3. Price tags sometimes get so sticky that they can ruin any object. If you are considering to make a purchase, you may want to ask the vendor to carefully remove his price tag so that the exact condition is disclosed.

Tip 4. Sunlight can cause **permanent** discoloration on plastic or plush materials if an object is exposed over a significant period of time.

Tip 5. If an item is made up of multiple parts, take the time to make sure that all pieces are accounted for. There is nothing more disappointing then putting together a puzzle that has pieces missing.

Keep in mind that if a slight imperfection is in an inconspicuous place, then the object may still look nice on a display.

Most of us are on a budget; so every penny counts. Once you have established what type of condition the object is in, you should also estimate a fair price. Lately, it appears as though ceramic items in particular are commanding extremely high prices. I usually rate the object first on rarity, then on condition. If an object is rare, I would probably buy it at any price even with slight imperfections. However, if an object is common, I may wait until I can find it somewhere in mint condition.

I estimate the condition of an object on a scale from 1 to 10 and price it accordingly:

Grade	Condition	Definition	Price example
1-4	Poor	Item shows major signs of wear. No tags and/or packaging present. i.e., Noticeably broken, reglued, or discolored.	$1.00 - $4.00
5	Fair	Item shows minor signs of wear and has some missing components. No tags and/or packaging present. i.e., Missing puzzle or game pieces.	$5.00
6	Good	Item has been used and shows minor signs of wear. No tags and/or packaging present. i.e., Small chip in an inconspicuous area or very slight crazing.	$6.00
7	Very Good	Item has been used but is still pristine. No tags and/or packaging present.	$7.00
8	Excellent	Item has been gently used but is still pristine. Original tags and/or packaging may be ripped or damaged.	$8.00
9	Near mint	Item has been gently used but is still pristine. Original tags and/or packaging may be unsealed and slightly worn.	$9.00
10	Mint	Item has never been used and is pristine. Original tags are present and/or packaging is sealed from the factory.	$10.00

One last thing. Vendors at flea markets and toy shows **expect** you to barter on prices. Do not be shy! If an object is priced too expensive for the condition that it is in, let the vendor know in a diplomatic manner. If your criticism is warranted, the vendor will often reduce his asking price. Best of luck in your hunting endeavors!

Chapter One
Figurines

Figurines, ceramic. Enesco Corporation. Complete set of five. Garfield 2", Odie 3", Arlene 2.25", Nermal 1.25", and Pooky 1.5". $20 each or $125 set.

Figurine, ceramic. Enesco Corporation. Garfield is dressed in a pumpkin costume. 3.25". $20-35.

Figurine, ceramic. Enesco Corporation. Garfield is lying on top of a turkey. 2.5". $20-35.

Figurine, ceramic. Enesco Corporation. Garfield is skiing. 3". $20-35.

Figurine, ceramic. Enesco Corporation. Garfield is jogging. 2.5". $20-35.

Figurine, ceramic. Enesco Corporation. Garfield is playing basketball. 2.5". $20-35.

Figurine, ceramic. Enesco Corporation. Garfield is playing baseball. 2.5". $25-40.

Figurine, ceramic. Enesco Corporation. Garfield is playing hockey. 3". $20-35.

Figurine, ceramic. Enesco Corporation. Garfield is playing football. 2.5". $20-35.

Figurine, ceramic. Enesco Corporation. Garfield is roller skating. 2.5". $20-35.

Figurine, ceramic. Enesco Corporation. Garfield is playing tennis. 2.5". $20-35.

Figurine, ceramic. Enesco Corporation. Garfield is sitting. 4". $40-55.

Figurine, ceramic. Enesco Corporation. Garfield is grinning. 4". $40-55.

Figurine, ceramic. Enesco Corporation. Odie is sitting in a basket full of Easter eggs. The handle of the basket is wicker. 5.25". $45-60.

Figurine, plastic. Decopac/Russ. Garfield is holding a microphone. 3". $10-15.

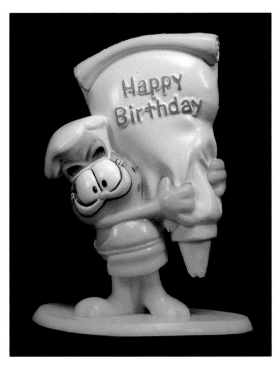

Figurine, plastic. Decopac/Russ. Garfield is holding an oversized icing bag. Caption: "Happy birthday." 3.5". $10-15.

Figurine, plastic. McDonald's promotion (under age 3), late 1980s. Garfield is wearing a headband, elbow/knee pads, and skates. 2.5". $10-15.

Figurine, plastic. McDonald's promotion (under age 3), late 1980s. Garfield is hugging Pooky while skateboarding. 2.75". $10-15.

Figurine, plastic. R. Dakin & Company. Garfield is playing basketball. 2.75". $5-10.

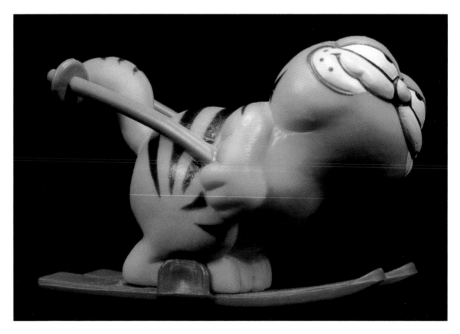

Figurine, plastic. Manufacturer unknown. Garfield is skiing. 1.25". $5-10.

Figurine, plastic. Manufacturer unknown. Garfield is golfing. 2.25". $5-10.

Figurine, plastic. Manufacturer unknown. Garfield is playing soccer. 2.25". $5-10.

Figurine, plastic. Manufacturer unknown. Garfield is jogging. 2.25". $5-10.

Figurine, plastic. Manufacturer unknown. Garfield is playing tennis. 2.25". $5-10.

Figurine, plastic. Manufacturer unknown. Garfield is playing baseball. 2.5". $10-15.

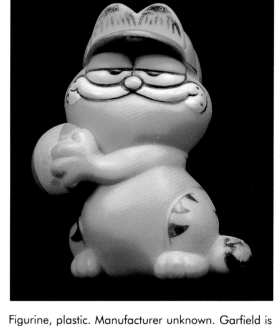

Figurine, plastic. Manufacturer unknown. Garfield is playing with a beach ball. 2.25". $5-10.

Figurine, plastic. Dakin, Inc. Another version of Garfield playing baseball. 2.5". $10-15.

Figurine, plastic. Manufacturer unknown. Garfield is sitting. 1.75". $5-10.

Figurine, plastic. Manufacturer unknown. Garfield is sleeping on a pillow. 1.5". $5-10.

Figurine, plastic. Manufacturer unknown. Odie is shown in two different poses. Sitting up 2.5" and laying down 1.5". $10-15 each.

Figurine, plastic. Manufacturer unknown. Garfield is wearing a fur coat. 2". $5-10.

Figurine, plastic. Bully. Garfield is opening a bottle of champagne. 3". $10-15.

Figurine, rubber. Dakin, Inc. Bendable. Various quotes were placed on belly. For example: "Am I cool or what?" or "Take me, I'm yours!" 4.5". $15-20.

Figurine, plastic. Enesco Corporation. Various quotes were placed on hearts. For example: "Kiss me," "I love you," and "Love me." 1.75". $20-35.

Figurine, plastic. Manufacturer unknown. Stackable. Various quotes were placed on signs. For example: "Hang in there!" or "Loves me!" 4.5". $15-20 each.

Figurine, plastic. Manufacturer unknown. Stackable. Garfield. 2.25". $5-10 each.

Figurine, plastic. Enesco Corporation. Garfield is dressed in green for St. Patrick's day. Titled: "Full -o- Blarney." 6.75". $20-35.

Chapter two
Trophies

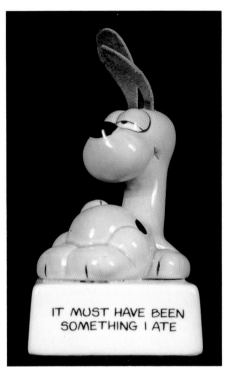

Trophy, ceramic. Enesco Corporation. An overstuffed green Odie is shown. Caption: "It must have been something I ate." 3.75". $40-55.

Trophy, ceramic. Enesco Corporation. Garfield is looking at a picture of himself. Caption: "I think I'm in love!" 3.75". $35-50.

Trophy, ceramic. Enesco Corporation. Garfield is hugging Pooky. Caption: "Happy birthday." 4.5". $35-50.

Trophy, ceramic. Enesco Corporation. Garfield is holding a cake. Caption: "Blow out the candle, and let's eat." 3.75". $35-50.

Trophy, ceramic. Enesco Corporation. Garfield is wearing a graduation cap and gown. Caption: "Look out world...here I come." 3.75". $35-50.

Trophy, ceramic. Enesco Corporation. Garfield is holding an apple. Caption: "Look, we all can use brownie points." 3.5". $35-50.

Trophy, ceramic. Enesco Corporation. Garfield has a cast on his leg and is on crutches. Caption: "Klutz!" 4". $35-50.

Trophy, ceramic. Enesco Corporation. Garfield is in a bunny outfit. Caption: "Here comes Kitty Cottontail." 3.5". $35-50.

Trophy, ceramic. Enesco Corporation. Garfield is laying down. Caption: "Eat and sleep. Eat and sleep. There must be more to a cat's life, but I hope not." 2.5". $35-50.

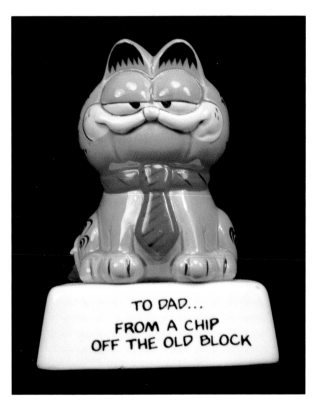

Trophy, ceramic. Enesco Corporation. Garfield is wearing a striped tie. Caption: "To Dad...from a chip off the old block." 4". $35-50.

Trophy, ceramic. Enesco Corporation. Odie is standing with his tongue hanging out. Caption: "Arf!" 4.75". $40-55.

Trophy, ceramic. Enesco Corporation. Garfield is sitting at a desk. Caption: "I'm so happy here I could just barf." 3.75". $70-85.

Trophy, ceramic. Enesco Corporation. Garfield is dressed in a green hat, coat, and bow tie. Caption: "We're all Irish one day a year." 3". $35-50.

Trophy, ceramic. Enesco Corporation. Garfield is on a pedestal. Caption: "Life is arriving at a mutually acceptable compromise...you give me the munchy and I'll let you keep your face." 5". $35-50.

Trophy, ceramic. Enesco Corporation. Odie is popping out of a trophy. Caption: "I love your bones." 6.25". $40-55.

Trophy, ceramic. Enesco Corporation. Garfield is holding a heart which says, "Be my Valentine." Caption: "I'm all yours." 3". $35-50.

Trophy, ceramic. Enesco Corporation. Garfield is leaning over towards Arlene. Caption: "Where have you been all my life?" 4.25". $50-65.

Trophy, plastic. Enesco Corporation. Garfield is dressed as a physician. Caption: "One X-ray is worth a thousand dollars." 3.75". $30-45.

Trophy, ceramic. Enesco Corporation. Garfield and Odie are dressed up as Laurel and Hardy. Caption: "This is another fine mess you've gotten us into." 4.5". $50-65.

Trophy, plastic. Enesco Corporation. Garfield is dressed like a teacher. Caption: "I need substitute students." 3.25". $30-45.

Trophy, plastic. Enesco Corporation. Garfield is sitting in a punch bowl. Caption: "The party's over!" 3.5". $30-45.

Trophy, plastic. Aviva. Garfield is kicking a football. Caption: "When in doubt punt." 4". $10-15.

Trophy, plastic. Enesco Corporation. Garfield is surfing in a bird bath. Caption: "Life's a beach party." 4.25". $30-45.

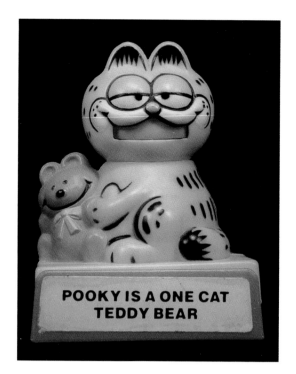

Trophy, plastic. Manufacturer unknown. When you push down on the trophy base, Garfield's head will move back and forth. Various quotes were placed on the base label. For example: "Big fat hairy deal." or "Take me home feed me!" 3.5". $15-20.

Trophy, plastic. Manufacturer unknown. When you push down on the trophy base, Garfield's teeth will show. Various quotes were placed on the base label. For example: "Pooky is a one cat teddy bear." or "I never met a lasagna I didn't like." 3.5". $15-20.

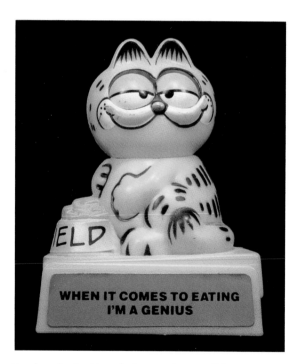

Trophy, plastic. Manufacturer unknown. When you push down on the trophy base, Garfield's eyes will open and close. Various quotes were placed on the base label. For example: "When it comes to eating, I'm a genius." or "I hate cute." 3.5". $15-20.

Snowdome, plastic. Enesco Corporation. Garfield is proudly displaying his diploma. Caption: "Congratulations!" 3.5". $15-20.

Chapter three
Kitchen

Cookie jar, ceramic. Enesco Corporation. Garfield is contently rubbing his tummy. 11.25".
$150-165.

Cookie cutters, plastic. Wilton Enterprises, Inc. Four piece set including: Garfield laying down, Odie laying down, Garfield sitting, and Garfield's head. $5-10 each.

Baking cups. Wilton Enterprises, Inc. The pattern on the cup shows Garfield jumping out of a cake. 50 Baking and party cups. Standard size. $5-10.

Cake decorations, plastic. Wilton Enterprises, Inc. Four piece set containing: Garfield, Odie, and Pooky with a wagon. $20-35.

Orange juice container, glass. Manufacturer unknown. Garfield is grinning with orange slices covering his eyes. 8.5". $10-15.

Creamer, ceramic. Enesco Corporation. Garfield's paw is the spout and his tail is the handle. 5.5". $85-100.

Bag clips, plastic. Guardsman Products, Inc. Garfield is shown pouncing. $5-10

Magnet, plastic. Enesco Corporation. Garfield is clinging to a shamrock. Caption: "Stuck on the Irish." $10-15.

Bag clips, plastic. Guardsman Products, Inc. Garfield's head. $5-10.

Magnet, plastic. Manufacturer unknown. Garfield has miniature suction cups on each paw. $10-15.

Coupon holder, vinyl. Ambassador. Garfield is shown waving. Caption: "A coupon saved is a coupon earned." $5-10.

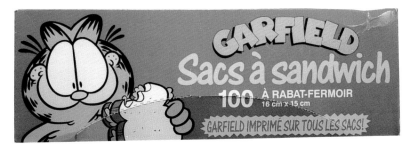

Sandwich bags. Northern Plastics, Ltd. An orange outline of Garfield is printed throughout the clear bag. 100 bags per box. $5-10.

Foodwrap. Northern Plastics, Ltd. An orange outline of Garfield is printed throughout the wrap. Microwaveable. $5-10.

Trivet, ceramic. Enesco Corporation. Garfield is dressed in a blazer and pin striped pants. Caption: "I was just another cat until I went to school." 4.25." $20-35.

Trivet, ceramic. Enesco Corporation. Garfield balances himself on Pooky while reaching for a pie. Caption: "I like you because...you help me reach my goals. 4.25". $20-35.

Lunch bags, paper. Carrousel Party Favors, Inc. A playful Odie is pictured. $5-10.

Air Freshener. Marlenn. Available in various shapes. Garfield is shown with a full dish of food. $5-10.

Lunchbox, plastic. Thermos. Garfield is shown with his arms outstretched against a festive background. $20-35.

Pumpkin carving kit, plastic. Pumpkin Masters, Inc. Kit includes eight designs; one of which is Garfield. Some of the original carving tools are not shown. $5-10.

Lunchbox, plastic. Thermos. Odie is giving Garfield a big kiss. $20-35.

Lunchbag, insulated canvas. Thermos. Garfield is putting a puzzle together. The bag is tan with red accents. $25-40.

Fruit snack box. General Mills, Inc. Chewy fruit snacks in the shape of Garfield and friends. $15-20.

Lollipop mold, rubber. Manufacturer unknown. Garfield. $5-10.

Chocolate. Allan. A milk chocolate shaped Garfield. $5-10.

Candy container, plastic. Superior Toy. Garfield is holding a red container for gumballs. 3". $5-10.

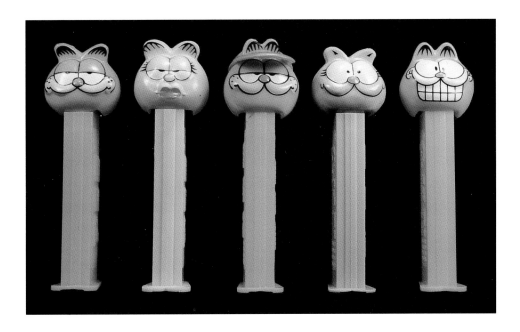

Candy container, plastic. Pez Candy, Inc., late 1980s. Five in series. Garfield, Arlene, Garfield wearing a green visor, Nermal, and Garfield Grinning with teeth. $5-10 each.

Candy container, plastic. Pez Candy, Inc., late 1990s. Four in series. Garfield pilot, Odie, Garfield cook, and Lazy Garfield. $5-10 each.

Candy container, rubber/plastic. Bee International, late 1990s. Garfield is peeking his head out of a present. Candy varied. $5-10.

Pizza box. Pizza Hut promotion, early 1990s. Garfield is licking his fingers. Caption: "Your pizza was -burp- delicious." $10-15.

Can. Campbell Soup Company/Franco American, early 1990s. Garfield is on a SpaghettiOs label juggling hamburgers. Garfield PizzaOs. $15-20.

Cat food box. Alpo Petfoods, Inc. promotion, early 1990s. Garfield is on the package of Alpo Gourmet Dinner cat food. $10-15.

Can. Campbell Soup Company/Franco American, early 1990s. Garfield is on a SpaghettiOs label holding a piece of pizza. Garfield PizzaOs. $15-20.

34

Chapter four
Mugs and Cups

Mug, glass. McDonald's promotion, late 1980s. Garfield is on a skateboard that Odie is pulling. Caption: "Use your friends wisely." $10-15.

Mug, glass. McDonald's promotion, late 1980s. Garfield is riding a see-saw with his friends. Caption: "I'm not one who rises to the occasion." $10-15.

Mug, glass. McDonald's promotion, late 1980s. Garfield and Odie are paddling in different directions. Caption: "I'm easy to get along with when things go my way." $10-15.

Mug, glass. McDonald's promotion, late 1980s. Garfield is lounging in a hammock while Odie is holding up one end. Caption: "It's not a pretty life but somebody has to live it." $10-15.

Mug, glass. McDonald's promotion, late 1980s. Garfield is sleeping. Caption: "I've never seen a sunrise...I'm waiting for the movie." $10-15.

Mug, ceramic. Enesco Corporation. The handle is shaped like Garfield. Caption: "I resent that." $30-55

Mug, ceramic. Enesco Corporation. Garfield is dipping his paw into a fish tank. Caption: "Call it an ethnic weakness." $20-35.

Mug, ceramic. Enesco Corporation. Caption: "I love the Irish." $5-10.

Mug, ceramic. Enesco Corporation/NFL Officially Licensed Product. Caption: "I'm a BILLS fan-atic!" $5-10.

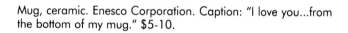

Mug, ceramic. Enesco Corporation. Caption: "I love you...from the bottom of my mug." $5-10.

Mug, ceramic. Enesco Corporation. Caption: "Garfield... last of the red hot lovers." Heart shaped handle. $10-15.

Mug, ceramic. Enesco Corporation. Caption: "Hugs & kisses." $5-10.

Mug, ceramic. Enesco Corporation. Caption: "I like you be-cause... you're so funny." $5-10.

Mug, ceramic. Enesco Corporation. Caption: "I like you because...you're so soft." $5-10.

Mug, ceramic. Enesco Corporation. Caption: "Cool daddy." $5-10.

Mug, ceramic. Enesco Corporation. Two interlocking mugs. Arlene and Garfield are pictured kissing. Heart shaped handles. $20-35 set.

Mug, ceramic. Enesco Corporation. Caption: "I finally did it!" and "Ha! Made you look." $5-10.

Cup, plastic. Pizza Hut promotion, early 1990s. Air Garfield series. Set of four. Garfield as an airplane pilot and Odie as an astronaut are shown. $5-10 each.

Chapter five
Bath

Cup holder, plastic. Manufacturer unknown. Garfield is wearing a bow tie on the dispenser. The cups have various pictures and sayings. Captions: "Parties are my life!" or "Looks like the moon is in a party phase." $15-20.

Adhesive bandages. Kid Care. Thirty strips per package. A pattern of Garfield's head is against a blue background. $5-10.

Tissues. Kid Care. Multiple pictures of Garfield and friends are shown on three 2-ply packs. $5-10.

Nail brush, plastic. Kid Care. Garfield is laying down with bristles underneath him. $5-10.

Liquid soap. Kid Care. The shape of the container is Garfield. $15-20.

Lip balm. Kid Care. A picture of Garfield laying down is on the blue and orange container. $5-10.

Bubble bath and shampoo, plastic. Menley & James Laboratories, Inc. The bubble bath container has a Magic Label picture that appears when it is submerged in warm water. $15-20 each.

Soap holder, vinyl. Avon Products, Inc. Garfield is made into an inflatable holder for soap. $10-15.

Soap holder, plastic. Kid Care. Garfield and friends are on a yellow container. $5-10.

Bar soap. Avon Products, Inc. Garfield is on the packaging wearing scuba gear. $5-10.

Mirror, plastic. Kid Care. Garfield is wearing a bow tie and is on a yellow container. $5-10.

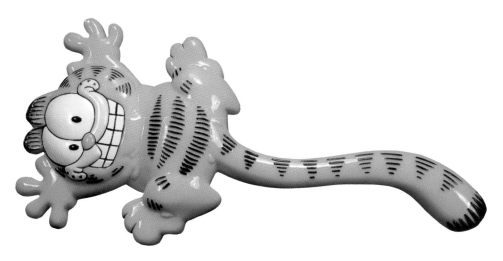

Bath brush, plastic. Avon Products, Inc. The bristles are tucked under Garfield's belly. $20-35.

Hair brush, plastic. Avon Products, Inc. The bristles are tucked under Garfield's belly. A comb was included but is not shown. $15-20.

Toothbrush holder, plastic. Avon Products, Inc. A child's toothbrush was incuded; however, the one pictured is not original. $20-35.

Toothbrush. John O. Butler Co.
Glow in the dark child's toothbrush.
$5-10.

Underwear box. Jockey. Both men's fly front briefs and tapered
boxers were produced. $15-20.

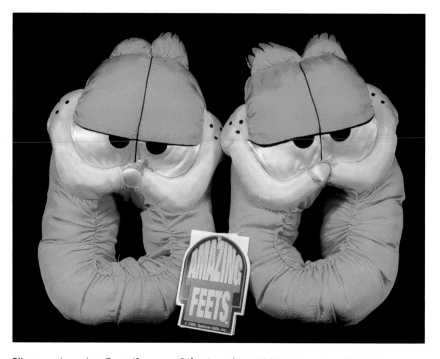

Slippers. Amazing Feets/Spencer Gifts, Inc., late 1990s. One size. $15-20.

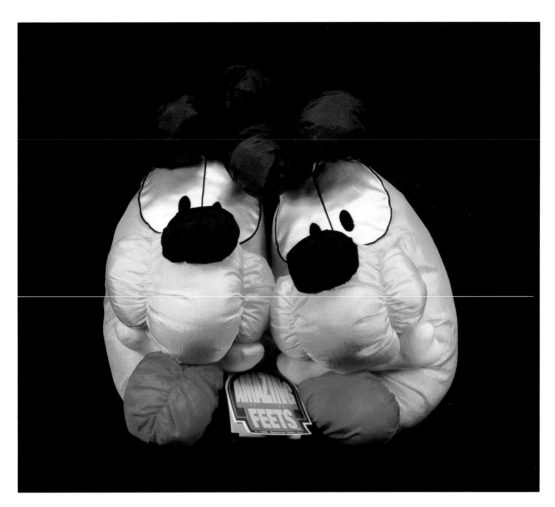

Slippers. Amazing Feets/Spencer Gifts, Inc., late 1990s. One size. $15-20.

Slippers, plush. Earthwood, Ltd. Garfield is shown comfortably tucked in bed. One size. $15-20.

47

Chapter six
General household

Telephone, plastic. Tyco Industries, Inc. The push button telephone is programmed to greet you with one of eleven sayings each time it rings. For example: "I'm not getting it." or "Yoo Hoo." $110-125.

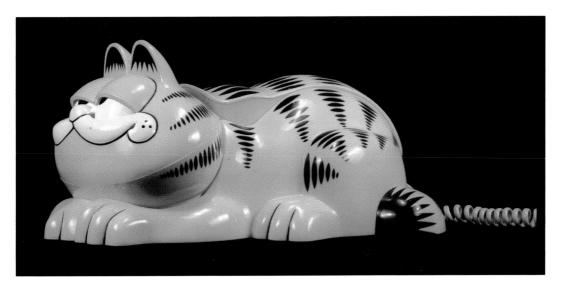

Telephone, plastic. Tyco Industries, Inc. Garfield's eyes open and close when you pick up the push button receiver that is camouflaged in his back. $85-100.

Telephone, plastic. Kash "N" Gold, Ltd. Telemania, late 1990s. Garfield's eyes open and close when you pick up the phone. The push buttons are on his belly. $65-80.

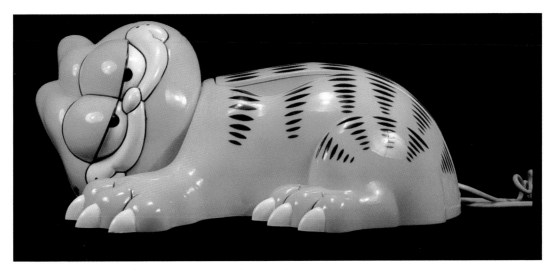

Telephone, plastic. Tyco Industries, Inc. Garfield's head rotates to each side. There is a push button receiver camouflaged in his back. $85-100.

Appliqués. 3M/Post-it notes. Complete with 29 pieces of self-stick removable appliqués. Includes a growth chart. $20-35.

Aquarium, plastic. Hawkeye Corporation. Garfield guards over his 2-gallon clear Lexan tummy tank. The kit includes an air pump, lighting, filtering system, fish food, and a dechlorinator. $110-125.

Therometer, plastic. Springfield. $15-20.

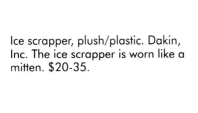

Ice scrapper, plush/plastic. Dakin, Inc. The ice scrapper is worn like a mitten. $20-35.

50

Radio, plastic. Durham Industries, Inc.
Caption: "Music is my life." Headphones
were included but are not pictured.
$35-50.

Alarm clock, metal. Sunbeam. Garfield is shown
laying down. $50-65.

Alarm clock, plastic. Sunbeam. Garfield's smile proudly
displays the digital time in bright red. The snooze alarm is
activated by squeezing his ear. The alarm is turned off by
pushing his nose. $50-65.

Wall clock, plastic.
Sunbeam.
Garfield's eyes
move back and
forth in synch with
his tail. $70-85.

51

Travel clock, plastic. Sunbeam. LCD quartz. $20-35.

Trinket box, ceramic. Manufacturer unknown. Top view shown. This container has a red lid and a white base. .75" in diameter. $15-20.

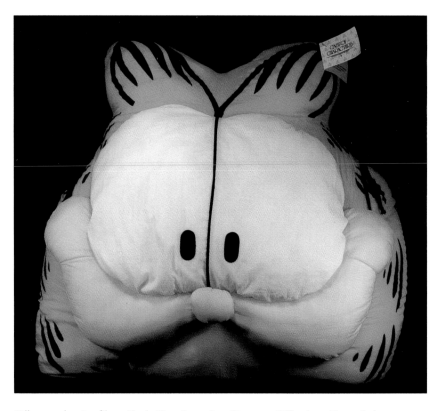

Pillow, polyester fiber. Plush Creations, Inc./Spencer Gifts, Inc. Cast of characters. Garfield's head. $40-55.

Wastebasket, metal. Cheinco. The reverse side shows four panels of comics. $50-65.

Sign, metal. Manufacturer unknown, late 1990s. Garfield crossing sign. $20-35.

Sign, metal. Manufacturer unknown, late 1990s. Garfield stop sign. $20-35.

Camera. 3M, late 1990s. Single use 35MM film and camera with flash. Caption: "Say lasagna." $10-15.

Massager, plastic. Pollenex. The Rival Company, late 1990s. $20-35.

Frame, plastic. Enesco Corporation. Garfield is wearing sunglasses. Caption: "Here's looking at you." $15-20.

Keychain, rubber. Manufacturer unknown. Garfield is standing. $5-10.

License plate, metal. Bureau of Motor Vehicles special sales/dealer, late 1990s. Caption: "Indiana committed to education." $10-15.

Candle. Enesco Corporation. Caption: "Love me." $5-10.

Candle. Manufacturer unknown. Caption: "I love you this much." $5-10.

Candle. Enesco Corporation. Garfield is sitting. $5-10.

Container, metal. Cheinco. Top view shown. This container has a pink lid with a white bottom. Caption: "Purrr." $10-15.

55

Container, metal. Cheinco. Caption: "Lighthearted." $10-15.

Container, metal. Dakin, Inc. Caption on lid: "I want out!" $10-15.

Container, metal. Cheinco. Garfield is shown holding a can opener. $10-15.

Bag, canvas. Manufacturer unknown. Caption: "Feed me." One size. $5-10.

Life jacket. Ero Industries, Inc. Children's sizes. $20-35.

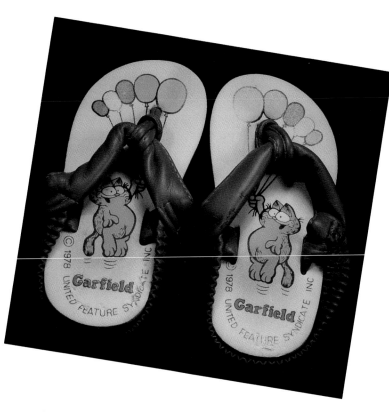

Flip Flops. Manufacturer unknown. Children's sizes. $5-10.

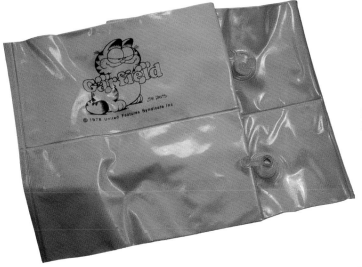

Flotation device. TAK Shing. These are inflatable and can be used to assist with swimming. Children's sizes. $5-10.

Cap. Just 4 Kids. Red and white striped cap. One size. $10-15.

Collector plate, ceramic. The Danbury Mint. Limited edition to 75 firing days. Titled: "I'll rise but I won't shine." 8" in diameter. $45-60.

Collector plate, ceramic. The Danbury Mint. Limited edition to 75 firing days. Titled: "And now for dessert." 8" in diameter. $45-60.

Collector plate, ceramic. The Danbury Mint. Limited edition to 75 firing days. Dear diary series. Titled: "I met a charming cat in the mirror this morning." 8" in diameter. $45-60.

Collector plate, ceramic. The Danbury Mint. Limited edition to 75 firing days. Dear diary series. Titled: "Sleep, the perfect exercise." 8" in diameter. $45-60.

Chapter eight
Music boxes

Music box, plastic. Enesco Corporation. Garfield birthday action musical. Garfield's head pops in and out of the top of the cake. $85-100.

Music box, plastic. Enesco Corporation. Titled: "School spirit." Arlene is cheer leading while Odie tackles Garfield. Caption: "Football hero." $135-150.

Music box, ceramic. The Danbury Mint. The Garfield music box collection. Top view shown. The base is white. $100-115.

Chapter nine
Banks

Bank, ceramic. Enesco Corporation. Caption: "Donations accepted." 6". $70-85.

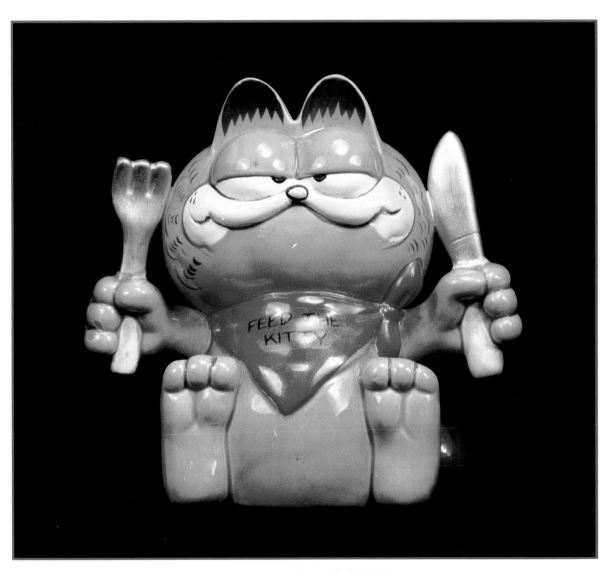

Bank, ceramic. Enesco Corporation. Caption: "Feed the kitty." 6". $70-85.

Bank, ceramic. Matscot International, L.L.C., late 1990s. Caption: "Think of all the money you could save for lasagna." 8.5". $30-45.

Gumball machine bank, plastic. Bee International, late 1990s. Garfield as a baseball player. 9.5". $15-20.

Gumball machine bank, plastic. Bee International, late 1990s. Garfield as a soccer player. 9.5". $10-15.

Gumball machine bank, plastic. Bee International, late 1990s. Garfield as a basketball player. 9.5". $10-15.

Bank, plastic. Bee International, late 1990s. Garfield as a football player. 9.5". $10-15.

Gumball machine bank, plastic. Superior Toy. Caption: "I'll share, but it will cost you!" Complete with gumballs, lock, and key. 6.5". $45-60.

Gumball machine bank, plastic. TimMee Toy. Garfield is perched on top. 8". $20-35.

Bank, glass. Manufacturer unknown. Garfield is sitting. 7". $15-20.

Chapter ten
Jewelry

Watch, metal/leather. Armitron, late 1990s. This watch came packaged in a plastic dome with a matching ceramic Garfield figurine. $85-100.

Watch, metal/leather. Armitron, late 1990s. This watch is shaped like Garfield's head. $30-45.

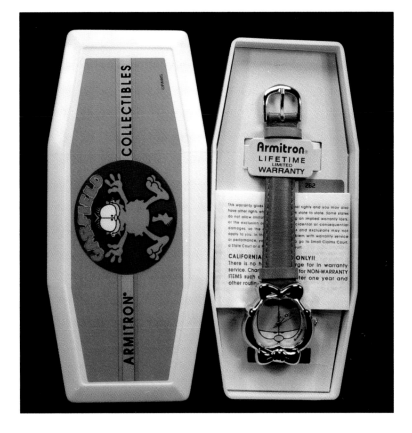

Pin, metal. Canadian McDonald's promotion (employees only), late 1980s. Caption: "Garfield collectibles, 59 cents each plus applicable tax while supplies last." 3.5" in diameter. $15-20.

Pin. McDonald's promotion (employees only), late 1980s. Garfield is sitting against the McDonald's golden arches logo. 1". $15-20.

Pin, metal. McDonald's promotion (employees only), late 1980s. Garfield is leaning against the McDonald's golden arches logo. 3.5" in diameter. $15-20.

Pin. McDonald's promotion (employees only), late 1980s. Garfield leaning against the McDonald's golden arches logo. 1". $15-20.

Pin, metal. Manufacturer unknown. Caption: "Diet is die with a T." 1.75" in diameter. $5-10.

Pin, metal. Manufacturer unknown. Caption: "I was just another cat until I went to school." 1.75" in diameter. $5-10.

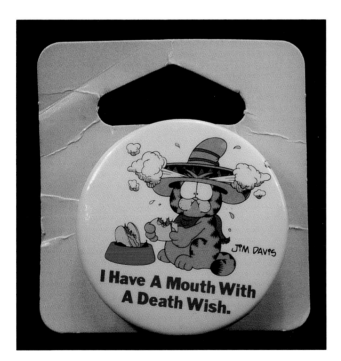

Pin, metal. Manufacturer unknown. Caption: "I have a mouth with a death wish." 1.75" in diameter. $5-10.

Pin, metal. Manufacturer unknown. Caption: "Party animal!" 2.25" in diameter. $5-10.

Pin, metal. Papersellers, Ltd. Caption: "Banzai!" 2.25" in diameter. $5-10.

Pin, plastic. Enesco Corporation/NFL Officially Licensed Product. Caption: "I'm a Bills fan-atic!" 3.25". $10-15. *Value increases between $20-35 if uniform reflects the former Cleveland Brown's logo.

Pin, plastic. Enesco Corporation. Caption: "You're the one for me." 3". $10-15.

Pin, plastic. Enesco Corporation. Caption: "I'm spooky." 1.5". $10-15.

Pin, plastic. Enesco Corporation. Caption: "I want you." 1.5". $10-15.

Chapter 11
Games and toys

Game, plastic. Parker Brothers. "Kitty letters" game. This game is composed of 40 sticks with letters on them. The object of the game is to spell a word without making the other sticks in the pile fall. $20-35.

Game, paper. Unique Industries, Inc. "Stick the tail on Garfield" game. Kit includes: 12 self-stick tails, game sheet, blindfold, and instructions. $15-20

Game. Parker Brothers. "Garfield" board game. $20-35.

Yo-yo, plastic. Avon Products, Inc. $10-15.

Puzzle, plastic. Unique Industries, Inc. Four sliding puzzles. $5-10 each.

Puzzle. Golden. Garfield is shown on a boat with Odie. 100 pieces. $5-10.

Puzzle. The Underground Jigsaw Puzzle. Caption: "Who knows where that mouse has been?" 100 pieces. $10-15.

Puzzle. Golden. Garfield is shown in a dorm room eating an entire pizza by himself. 200 pieces. $10-15.

Puzzle. Golden. Garfield is holding a nest filled with three blue eggs. 100 pieces. $5-10.

Puzzle. Golden. Garfield is wearing a space suit and is surrounded by heart-shaped creatures. $5-10.

Puzzle. MB Puzzle. Garfield and Odie are playing with a leaf. $5-10.

Puzzle. The Rainbow Works. Garfield is floating on a raft while holding a sandwich. 100 pieces. $5-10.

Puzzle. The Underground Jigsaw Puzzle. Caption: "This should hold you cat lovers for a while." 1000 pieces. $20-35.

Puzzle. Golden. Frame-Tray puzzle. Garfield is riding a unicycle while Odie runs along the side. $5-10.

Puzzle, wood. Playskool. Titled: "Hello, flower garden." Garfield is shown playing in a flower garden. 9 pieces. $10-15.

Activity kit. Landoll's, Inc. Includes five books, paint, and crayons. $20-35.

Coloring book. Landoll's, Inc. Titled: *Garfield The Sportin' Life.* $5-10.

Coloring and activity book. Landoll's, Inc. Titled: *Garfield Soaks Up Some Fun.* $5-10.

Sticker book. Landoll's, Inc. Titled: *Garfield Playful Pets.* $5-10.

Suncatcher. SunCatcher product from Makit & Bakit. Garfield is shown wearing a tux. $10-15.

Bucket, plastic. Berry Plastics, late 1990s. Subway promotion. Kid's Pak. Garfield, Nermal, and Pooky are shown playing tug of war with Odie, Arlene, and Squeak. The cover of the bucket is a 3-D image of Odie. $10-15.

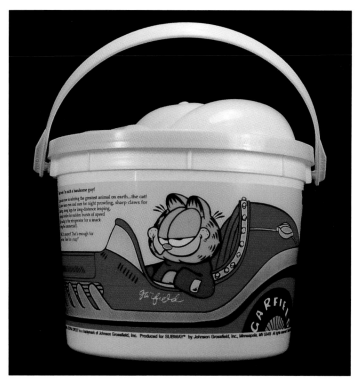

Bucket, plastic. Berry Plastics, late 1990s. Subway promotion. Kid's Pak. Garfield is shown in a red sports car. The cover of the bucket is a 3-D image of Odie. $10-15.

Trading cards. SkyBox International. Premier edition. Eight Garfield collector cards per pack. Tattoos and holograms were randomly packaged. $5-10 each pack. $10-15 each hologram. Promotional display box. $10-15.

Card games. The United States Playing Card Co. War card game. $5-10.

Card games. The United States Playing Card Co. Magic cards #1. $5-10.

Card games. The United States Playing Card Co. Crazy 8's. $5-10.

Card games. The United States Playing Card Co. Riddles & Jokes #1. $5-10.

View-Master reels. View-Master International Group. Titled: "City pound." $20-35.

Casting Kit, plastic. Berkley, mid 1990s. Purple and orange fishing pole complete with rod, casting plug, and line. Kit also included free stickers. $20-35.

Colorforms. Colorform Brand. Titled: "Garfield big birthday party play set." Kit contains plastic pieces and a play board. $25-40.

Toy, plastic/plush. Mattel. "Talking gourmet." Garfield's eyes open and close when he talks.   Once the pull string is activated, Garfield is programmed to say one of eight sayings. For example: "Take a cat to lunch." or "Stuffed again." $85-100.

Toy, metal. Ertl, early 1990s. Garfield is peddling an ice cream cart. Caption on original packaging: "Going my way?" $20-35.

Toy, metal. Ertl, early 1990s. Garfield is in a sports car. Caption on original packaging: "I'll drive." $20-35.

Toy, metal. Ertl, early 1990s. Garfield is driving a lasagna truck. Caption on original packaging: "Next stop, fun!" $20-35.

Toy, metal. Ertl, early 1990s. Odie is on top of a doghouse. Caption on original packaging: "Let's roll." $25-40.

Toy, metal. Ertl, early 1990s. Garfield is riding on a USA space shuttle. Caption on original packaging: "Need a lift?" $20-35.

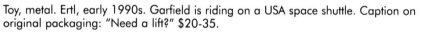

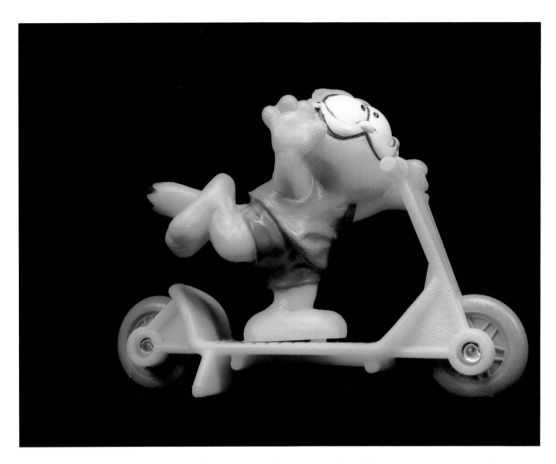

Toy, plastic. McDonald's promotion (first in series), late 1980s. Garfield is riding on a scooter. $5-10.

Toy, metal. Ertl, early 1990s. Garfield is dressed as a cowboy riding on a rocking horse. Caption: "Ride on!" $20-35.

Toy, plastic. McDonald's promotion (second in series), late 1980s. Garfield is driving a four wheeler. $5-10.

Toy, plastic. McDonald's promotion (third in series), late 1980s. Garfield is riding on a skateboard. $5-10.

Toy, plastic. McDonald's promotion (fourth in series), late 1980s. Garfield is driving a motorcycle. Odie co-pilots in the attached sidecar. $10-15.

Toy, plastic. Pizza Hut promotion, early 1990s. Titled: "Air Garfield paratrooper." $5-10.

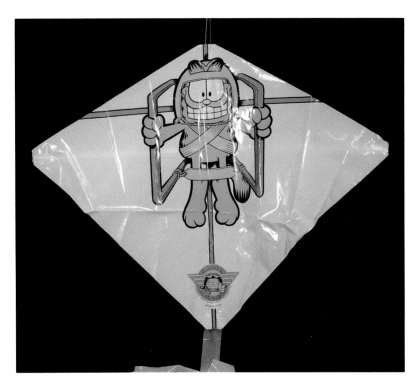

Toy, plastic. Pizza Hut promotion, early 1990s. Titled: "Air Garfield kite." $5-10.

Toy, vinyl. Pizza Hut promotion, early 1990s. Titled: "Air Garfield inflatable spaceball." $5-10.

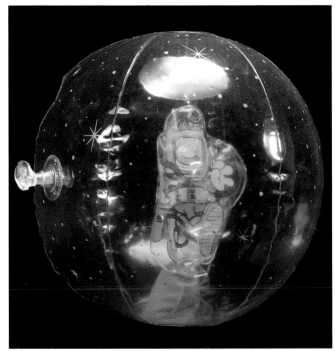

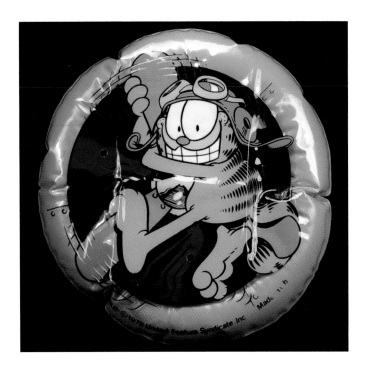

Toy, vinyl. Pizza Hut promotion, early 1990s. Titled: "Inflatable Garfield flyer." $5-10.

Toy, plastic. Playmates Toys, Inc. Garfield rides on top of the television set. $15-20.

Toy, plastic. NCM Corp. Bicycle reflector premium. $5-10.

Toy, plastic. Playmates Toys, Inc. When the pompom on the nightcap is wound, Garfield moves up and down. $15-20.

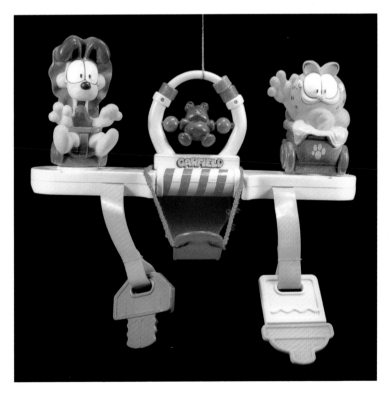

Toy, plastic. Remco Baby, Inc., early 1990s. This play center can be strapped to a stroller. $15-20.

Teether. Remco Baby, Inc. Soft teether for babies. $5-10.

Rattle, plastic. Remco Baby, Inc. Caption: "It's a girl!" $5-10.

Rattle, plastic. Remco Baby, Inc. Garfield hugs a circular rattle. $5-10.

Rattle, plastic. Remco Baby, Inc. "Diaper pin-pals." Garfield rattle. $5-10.

Rattle, plastic. Remco Baby, Inc. "Diaper pin-pals." Odie rattle. $5-10.

Rattle, plastic. Remco Baby, Inc. Garfield is driving a train. $5-10.

Toy, plastic. Remco Baby, Inc. Squeeze action. When the toy is squeezed, the balls circulate throughout the top of the plastic dome. $5-10.

Toy, plastic. Playmates Toys, Inc. Garfield shaped ball. $5-10.

Toy, plastic. Remco Baby, Inc. Squeeze pals. Garfield is wearing an oversized red nightshirt and bunny slippers. $5-10.

Puppet, plush. R. Dakin & Company. $20-35.

Toy, plush. R. Dakin & Company. Garfield clip-on. $10-15.

Toy, plush. R. Dakin & Company. Garfield clip-on.
$10-15.

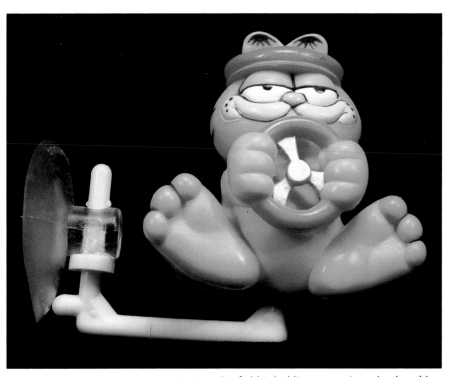

Toy, plastic. R. Dakin & Company. Suction. Garfield is holding a steering wheel and is
wearing a red cap. $10-15.

Toy, plush. R. Dakin & Company. Titled: "Garfield & Pooky." Garfield is shown giving Pooky a big bear hug. $40-55.

Toy, plush. R. Dakin & Company. Odie is standing with his tongue hanging out. $20-35.

Box. R. Dakin & Company. Various captions on each side of box: "There's a big advantage to having several of me around." or "Take me home...feed me." $15-20.

Toy, plush. R. Dakin & Company. Garfield is
sitting. $20-35.

Toy, plush. R. Dakin & Company. Titled: "Garfield on the town."
Garfield is wearing formal attire. $25-40.

Toy, plush. R. Dakin & Company. Titled: "Jitter Critter." Garfield has a vibrator tucked
into his tummy which makes him move in different directions. $50-65.

Toy, plush. R. Dakin & Company. Titled: "Love struck." Garfield has an arrow through his head. $20-35.

Toy, plush. R. Dakin & Company. Titled: "Eats out." Garfield is eating a hamburger. $25-40.

Toy, plush. R. Dakin & Company. Titled: "Born to party." Garfield is wearing a lamp shade on his head. $25-40.

Toy, plush. R. Dakin & Company. Garfield is dressed as a skier. $25-40.

Toy, plush. R. Dakin & Company. Garfield is on a sled. $25-40.

Toy, plush. R. Dakin & Company. Nermal. Hard to find. $30-45.

Toy, plush. R. Dakin & Company. Arlene.
Hard to find. $30-45.

Toy, plush. Spencer Gifts, Inc., late 1990s. Garfield is shown popping out of a cake.
$40-55.

Toy, plush. Fun Farm. Garfield is dressed as a baseball player. $20-35.

Toy, plush. R. Dakin & Company. Garfield is dressed as a tourist. $25-40.

Toy, plush. R. Dakin & Company. Titled: "Garfield bowler." His shirt reads, "Gutter ball Garfield." $25-40.

Toy, plush. Fun Farm/R. Dakin & Company. Garfield is dressed as a boxer. $25-40.

Toy, plush. R. Dakin & Company. Odie is dressed as a jester. $25-40.

Toy, plush. R. Dakin & Company. Garfield is dressed as a golfer. His shirt reads, "Fore!" $25-40.

Toy, plush. R. Dakin & Company. Garfield is dressed up for a pep rally. $25-40.

Toy, plush. R. Dakin & Company. Odie is dressed as the Easter Bunny. The carrot is held on his tongue with Velcro. $25-40.

Toy, plush. R. Dakin & Company. Garfield is ready to go swimming. Odie has been transformed into an intertube. $40-55.

Toy, plush. R. Dakin & Company. Titled: "Mountie stuck on you." This was issued to commemorate the 10th birthday celebration of Garfield. $40-55.

Toy, plush. R. Dakin & Company. Titled: "Garfield stuck on you II." Suction cup Garfield with a big toothy grin. $15-20.

Toy, plush. R. Dakin & Company. This photo shows an example of the reverse side of a suction cup Garfield. The sayings on the buttons varied. $15-20.

Toy, plush. R. Dakin & Company. Suction cup Garfield with a smile. $15-20.

Toy, plush. R. Dakin & Company. Suction cup Garfield with a closed mouth grin. $15-20.

Toy, plush. R. Dakin & Company. Titled: "Garfield I don't do windows." Suction cup Garfield is shown with a duster and apron. $25-40.

Toy, plush. Fun Farm. Garfield is shown with an appliquéd face, ears, and stripes. $20-35.

Chapter Twelve
Party Favors

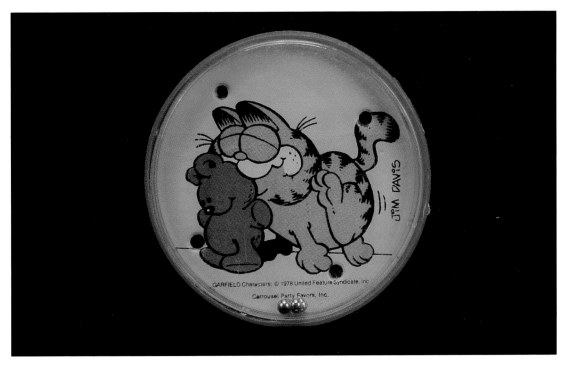

Game, plastic. Carrousel Party Favors, Inc. Garfield is shown cozying up to Pooky. $5-10.

Ring, plastic. Carrousel Party Favors, Inc. Top view shown. Ring flashes to show either Garfield brushing his teeth or with toothpaste all over him. $5-10.

Balloons. Accents/CTI. Six bright yellow balloons per package. A picture of Garfield smiling is printed on each balloon with red non-toxic ink. $5-10.

Streamers. Hallmark Cards, Inc. Garfield in various poses on a white background. $5-10.

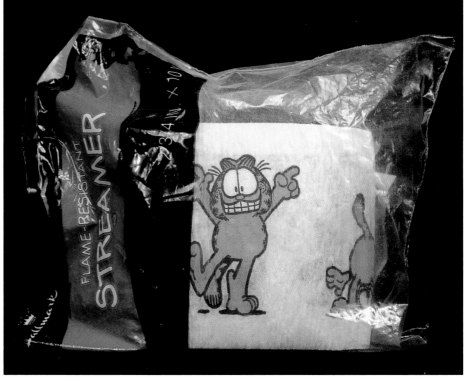

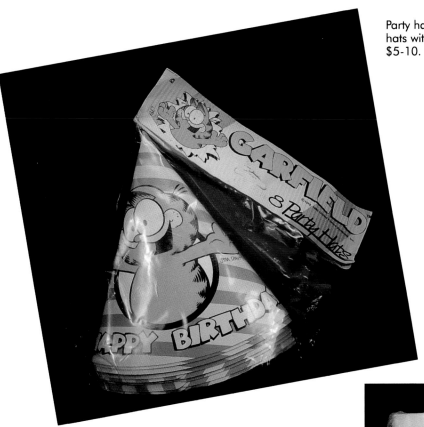

Party hats. Carrousel Party Favors, Inc. Pink and white striped hats with the caption: "Happy birthday." Eight hats per pack. $5-10.

Napkins. Gibson Greeting Cards, Inc. Caption: "Let's party." Sixteen napkins per pack. $5-10.

Napkins. Gibson Greeting Cards, Inc. Garfield standing. Sixteen napkins per pack. $5-10.

119

Loot bags. Unique Industries, Inc. Caption: "Ready to party when you are." Eight loot bags per pack. $5-10.

Squawkers. Unique Industries, Inc. Multi colored streamers attached to mouthpiece. $5-10.

Chapter Thirteen
Stationery

Notecards. Case Stationery Co., Inc. "Garfield visits
Norman Rockwell." Sixteen assorted note cards are inside
the reusable metal box. $20-35.

Postelettes. Stay Healthy card/American Health Foundation. Various designs were produced.
For example: Garfield working out, jogging, or looking in the mirror. $5-10.

Notes. 3M/Post-it. Self-stick removable notes. Various designs were produced. Pictured: Garfield grinning. $5.

Notes. 3M/Post-it. Self-stick removable notes. Various designs were produced. Caption: "It's been Monday all week." $5.

Stand-up cards, cardboard. Hallmark Cards, Inc. Various designs were produced. Caption: "My other body is young and athletic." $15-20.

Stand-up cards, cardboard. Hallmark Cards, Inc. Various designs were produced. Caption: "We're not cut out for a 9-to-5 job... maybe elevenish-to-two fifteen." $15-20.

Stand-up cards, cardboard. Hallmark Cards, Inc. Various designs were produced. Caption: "I gave up my diet for health reasons... I was sick of it." $15-20.

Puzzle. Gibson Greetings, Inc. Greeting card puzzle. Caption: "Be my Valentine." $15-20.

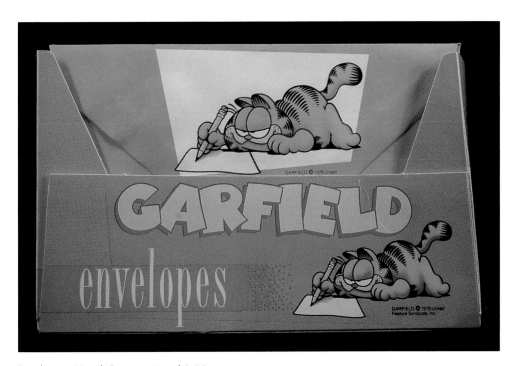

Envelopes. Mead Corporation. $5-10.

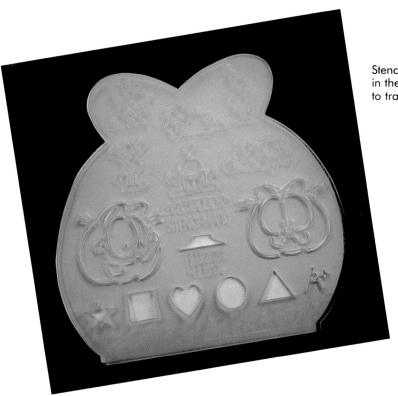

Stenci-matic, plastic. Pizza Hut promotion, early 1990s. A stencil in the shape of Garfield's head consists of cartoons and shapes to trace. $10-15.

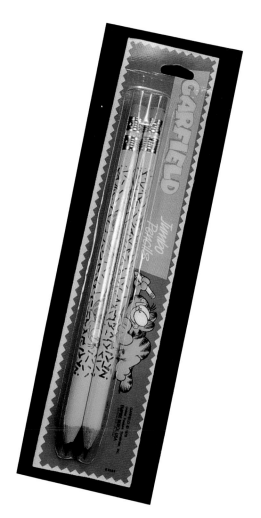

Pencil. Empire Berol USA. Two jumbo Garfield pencils. $5-10.

Stencil, plastic. Mead Corporation, late 1990s. A stencil in the full shape of Garfield. $5-10.

Pencil. Mead Corporation, late 1990s. Two jumbo Garfield pencils. $5-10.

Eraser. Mead Corporation, late 1990s. The eraser is in the shape of Garfield's head. $5-10.

Eraser. R. Dakin & Company. Garfield is shown holding Pooky. $5-10.

Pen. Mead Corporation, late 1990s. Multi-colored jumbo pen. $5-10.

Magnetic board, metal. Manufacturer unknown. Complete with paw shaped magnets. 4". $10-15.

Bulletin board. Young Things/R. Dakin & Company. Complete with bulletin board, stapler, staples, tacks, paper, colored pencils, eraser, and a Garfield pencil topper. 10". $15-20.

Glue stick. 3M/Scotch, late 1990s. Permanent adhesive. Non-toxic. Garfield is pictured on a purple background. $5-10.

Memo board. Antioch Publishing Co. Wipeable memo board with calendar. Caption: "Where did the weekend go?" 11". $10-15.

Book covers. Magic Cover division of Kittrich Corporation, late 1990s. Removable, self-adhesive vinyl. $5-10.

Tent cards. Enesco Corporation. Various designs were produced. Caption: "I'm all heart." $5-10.

Door knob hanger, paper. Manufacturer unknown. Various designs were produced. Pictured: "Enter at your own risk." and "Genius at work." $5-10.

Pencil sharpener, plastic. Manufacturer unknown. Garfield is laying down. $5-10.

Pencil sharpener/case, plastic accents. Empire Pencil Corporation. $15-20.

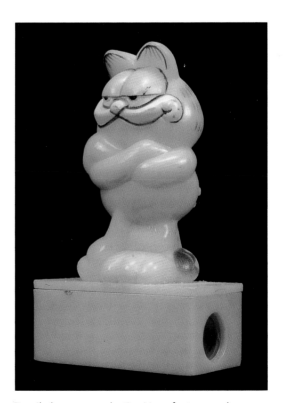

Pencil sharpener, plastic. Manufacturer unknown. Garfield is standing. $5-10.

Pencil toppers, plastic. Manufacturer unknown. Garfield smiling and climbing. $5-10 each.

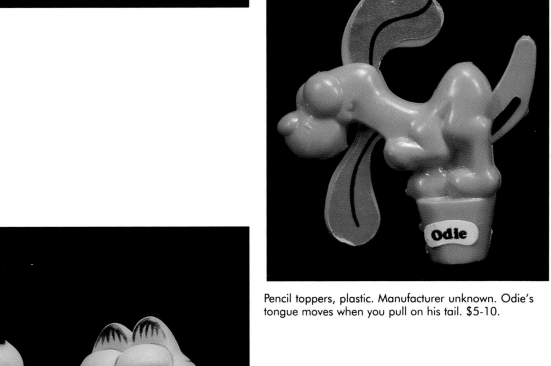

Pencil toppers, plastic. Manufacturer unknown. Odie's tongue moves when you pull on his tail. $5-10.

Pencil toppers, plastic. Manufacturer unknown. Odie and Garfield. $5-10 each.

Letter opener, plastic. Manufacturer unknown. Two different versions. Garfield's head is pictured on red background. Garfield is pictured leaning against a stack of papers on blue background. $5-10 each.

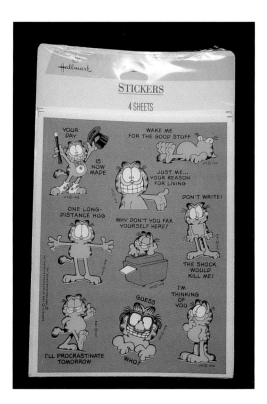

Stickers. Hallmark Cards, Inc. Garfield is shown in various poses. Captions include: "Your day is now made." and "Why don't you fax yourself here?" Four sheets. $5-10.

Stickers. Gibson Greetings, Inc. Poster stickers. Caption: "I think I'll go to class today...I need the sleep." One sheet. $10-15.

Stickers. Gibson Greetings, Inc. Captions include: "Let's be Valentines!" and "Be mine!" Three sheets. $10-15.

Sticker. Manufacturer unknown. Metallic sheen. Garfield is sitting on a fence with Arlene. Caption: "I think I'm in love." $10-15.

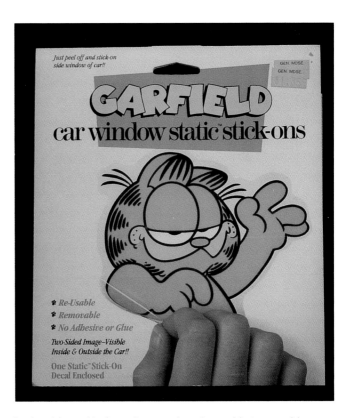

Static stickers. Marlenn Corporation. Re-usable/removable stickers. Garfield pictured. $5-10.

Static stickers. Marlenn Corporation. Re-usable/removable stickers. Odie pictured. $5-10.

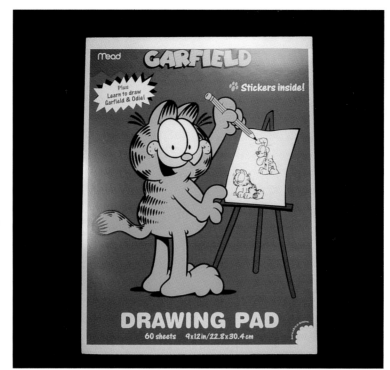

Drawing pad. Mead Corporation, late 1990s. Stickers and tips on how to draw Garfield and Odie are included. $5-10.

Rubber Stamp, plastic/rubber. Dakin, Inc. Garfield sits on top of a concealed ink pad. Caption: "Is it Friday yet?" $10-15.

Rubber stamp kit, foam/rubber. Rubber Stampede, late 1990s. Various designs: Garfield with a cup of coffee, Odie standing, and Garfield's head. Kit includes three stamps, ink pad, and sticker labels. $15-20.

Rubber Stamp, wood/rubber. Rubber Stampede, late 1990s. Various designs were produced. Pictured: Odie and Garfield with Pooky. $5-10 each.

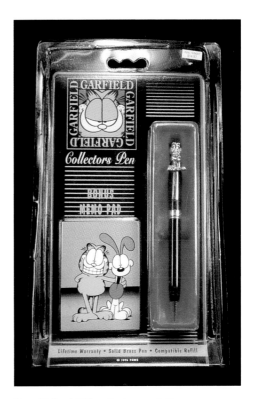

Pen. Stylus Writing Instrument Company. "Garfield collector's pen" with memo pad. $20-35.

Bookmark. Antioch Publishing Company. Various designs were produced. Pictured: "Keep your paws offa my stuff." "I'm so bright... I gotta wear shades." Garfield with heart-shaped balloons. $5-10 each.

Book, soft cover. A Golden Book, 1988. *Garfield the Big Star* created by Jim Davis. Story by Norma Simone. $5-10.

Book, soft cover. A Golden Book, 1990. *Mini-Mysteries featuring Garfield* by Jim Kraft. Illustrated by PAWS, Inc. $5-10.

Book, soft cover. A Little Golden Book, 1988. *Garfield and the Space Cat* story by Leslie McGuire. Created by Jim Davis. $5-10.

Book, soft cover. Random House, 1983. *Garfield Goes Under-ground* created by Jim Davis. Illustrated by Mike Fentz and Dave Kuhn. $5-10.

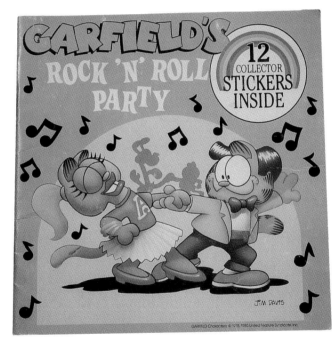

Book, soft cover. Antioch Publishing Co., 1989. *Garfield's Rock 'n' Roll Party* written by Jim Kraft. Design and illustration by: Dave Kuhn, Vicki Scott, and Mike Fentz. Twelve collector stickers were included. $5-10.

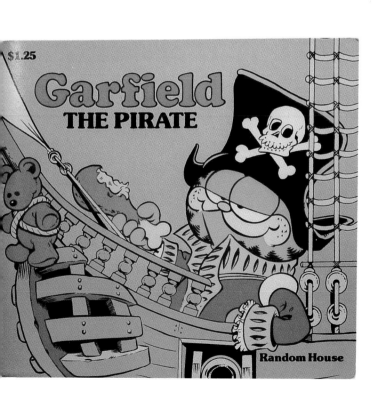

Book, soft cover. Random House, 1982. *Garfield the Pirate* created by Jim Davis. Illustrated by Mike Fentz. $5-10.

Book, soft cover. Antioch Publishing Co., 1991. *Garfield to the Rescue* written by Jim Kraft. Illustrated by Brett Koth and Larry Fentz, PAWS, Inc. Twelve collector stickers were included. $5-10.

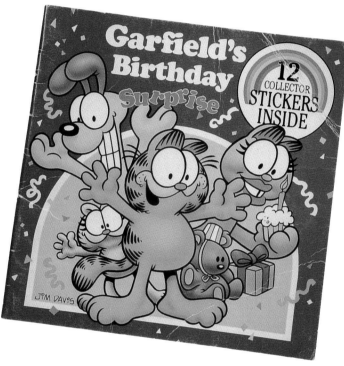

Book, soft cover. Antioch Publishing Co., 1987. *Garfield's Birthday Surprise*. Illustrations by Jim Davis. Twelve collector stickers were included. $5-10.

Book, hard cover. Grosset & Dunlap, Inc., 1989. *Garfield the Easter Bunny* created by Jim Davis. Written by Jim Kraft. $15-20.

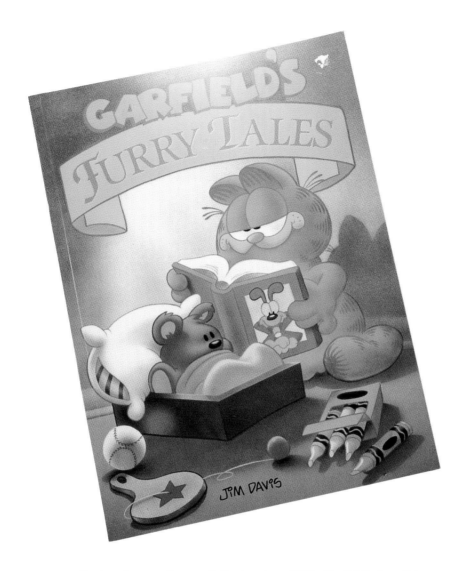

Book, soft cover. Grosset & Dunlap, Inc., 1989. *Garfield's Furry Tales* created by Jim Davis. Written by Mike Fentz. $5-10.

Book, hard cover. Western Publishing Company, Inc./A Golden Book, 1989 *Garfield and the Tiger* created by Jim Davis. Story by Jim Kraft. $15-20.

Chapter fifteen
Miscellaneous Christmas items

Ornament, ceramic. Enesco Corporation. Garfield is peeking out of a gift box wrapped in green and white striped paper. 3". $20-35.

Ornament, ceramic. Enesco Corporation. Garfield is hanging from a candy cane. 3.75". $20-35.

Ornament, plastic. Manufacturer unknown. Garfield is tangled up in a string of Christmas lights. 2.75". $30-45.

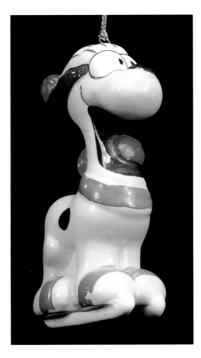

Ornament, ceramic. Enesco Corporation. Odie is shown on ice skates. 3.5". $25-40.

Ornament, plastic. Enesco Corporation, 1992-1994. Titled, "Holiday On Ice." Garfield is shown ice skating. 3.75". $30-45.

Ornament, plastic. Enesco Corporation, 1994. Titled, "Mine, Mine, Mine." Garfield guards a mailbox packed with gifts. 2.5". $30-45.

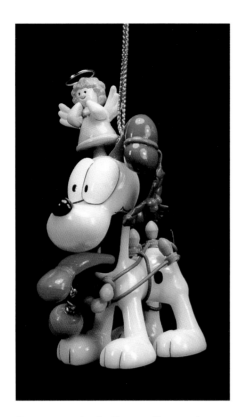

Ornament, plastic. Enesco Corporation, 1991-1992. Titled, "All Decked Out." Odie is shown tangled in a bunch of Christmas lights. 3". $35-50.

Ornament, plastic. Enesco Corporation. Odie is shown driving a red car which is tangled up in Christmas lights. The license plates reads, "Odie." 2.5". $35-50.

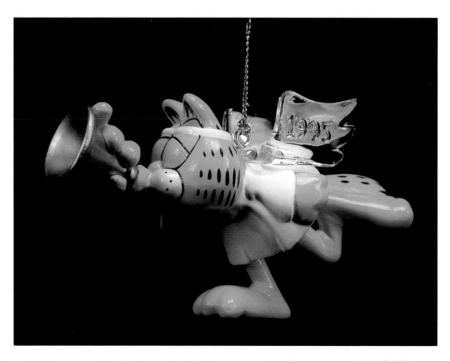

Ornament, plastic. Hallmark Cards, Inc., 1995. Keepsake ornament. Garfield is dressed up as an angel. 2.5". $30-45.

140

Ornament, plastic. Matrix Industries,
Ltd., late 1990s. Garfield's trim-a-tree
ornament. Garfield is wearing a Santa
hat while taking a nap on the moon. 3".
$20-35.

Ornament, plastic. Matrix Industries, Ltd., late 1990s. Garfield's
trim-a-tree ornament. Garfield is sitting in a sleigh that Odie is
pulling. 3". $40-55.

Ornament, plastic. Matrix Industries, Ltd., late
1990s. Garfield's trim-a-tree ornament. Garfield
is holding a purple box full of lights. 3". $20-35.

Ornament, plastic. Matrix Industries, Ltd., late 1990s. Garfield's
trim-a-tree ornament. Garfield is riding in a fire engine. 2.5".
$20-35.

Ornament, plastic. Matrix Industries, Ltd., late 1990s. Garfield's trim-a-tree ornament. Odie is pictured like a carousel horse. Garfield is riding on top. 3.5". $40-55.

Ornament, plastic. Matrix Industries, Ltd., late 1990s. Garfield's trim-a-tree ornament. Garfield is wearing an elf outfit and is holding Pooky. 2.5". $20-35.

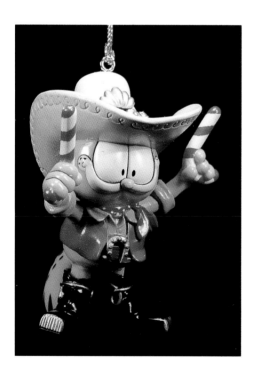

Ornament, plastic. Matrix Industries, Ltd., late 1990s. Garfield's trim-a-tree ornament. Garfield is wearing a cowboy outfit. 3.25". $20-35.

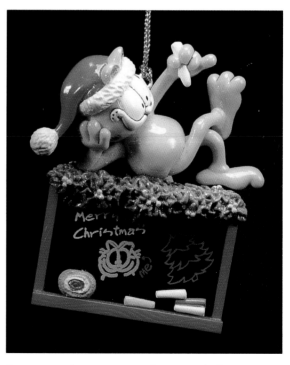

Ornament, plastic. Matrix Industries, Ltd., late 1990s. Garfield's trim-a-tree ornament. Garfield is sitting on top of a chalk board. 3". $20-35.

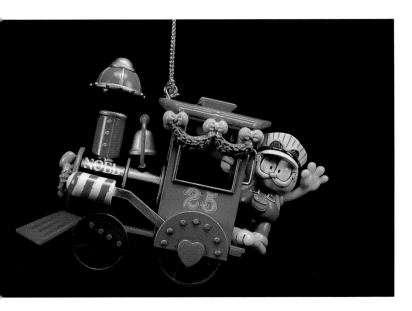

Ornament, plastic. Matrix Industries, Ltd., late 1990s. Garfield's trim-a-tree ornament. Garfield is riding on a train. 3". $20-35.

Ornament, plastic. Matrix Industries, Ltd., late 1990s. Garfield's trim-a-tree ornament. Garfield is shown with a stocking. 2.75". $20-35.

Ornament, plastic. Matrix Industries, Ltd., 1990s. Garfield's trim-a-tree ornament. Garfield is in a ballerina outfit. 3". $20-35.

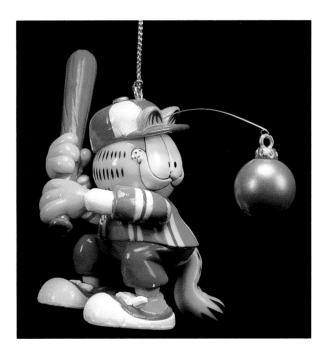

Ornament, plastic. Matrix Industries, Ltd., late 1990s. Garfield's trim-a-tree ornament. Garfield is playing baseball with an ornament. 3.25". $25-40.

Ornament, plastic. Matrix Industries, Ltd., late 1990s. Garfield's trim-a-tree ornament. Garfield is dressed as an ornament. 4". $20-35.

Ornament, plastic. Matrix Industries, Ltd., late 1990s. Garfield's trim-a-tree ornament. Garfield is peeking through a wreath. 3" in diameter. $20-35.

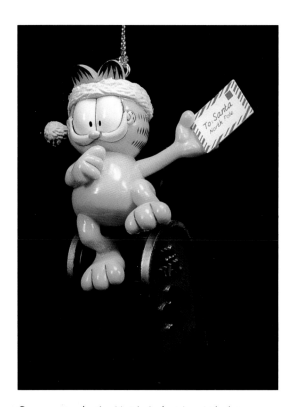

Ornament, plastic. Matrix Industries, Ltd., late 1990s. Garfield's trim-a-tree ornament. Garfield is mailing a letter to Santa. 3.75". $20-35.

Ornament, plastic. Matrix Industries, Ltd., late 1990s. Garfield's trim-a-tree ornament. Garfield is dressed as Santa and he is delivering some presents. 3". $20-35.

Ornament, plastic. Matrix Industries, Ltd., late 1990s. Garfield's trim-a-tree ornament. Garfield is skating on top of a popsicle. 3". $20-35.

Ornament, plastic. Matrix Industries, Ltd., late 1990s. Garfield's trim-a-tree ornament. Garfield is dressed as a football player. 3.25". $20-35.

Ornament, plastic. Matrix Industries, Ltd., late 1990s. Garfield's trim-a-tree ornament. Garfield is checking his shopping list. 3.5". $20-35.

Ornament, plastic. Enesco Corporation/NFL Officially Licensed Product. Garfield is shown wearing a New York Giants helmet. 2". $20-35. *Value increases between $50-65 if helmet reflects the former Cleveland Brown's logo.

Music box, plastic. Matrix Industries, Ltd., late 1990s. Garfield's sing-a-long collectible musical. Garfield is playing the piano. Arlene is dancing and Odie is howling. $135-150.

Bell, ceramic. Enesco Corporation, 1982. The bell's handle is in the shape of Garfield dressed as Santa. $40-55.

Bell, ceramic. Enesco Corporation, 1987. The bell is shaped like a Christmas tree; complete with Garfield dressed as an angel on top. $40-55.

Toy, plush. R. Dakin & Company. Suction cup Garfield wearing a Santa hat. $15-20.

Stocking hanger, plastic. Enesco Corporation. Garfield is tangled up in lights with an ornament in his hand. $50-65.

Toy, plush. R. Dakin & Company. Garfield clip-on wearing a Santa hat. $10-15.

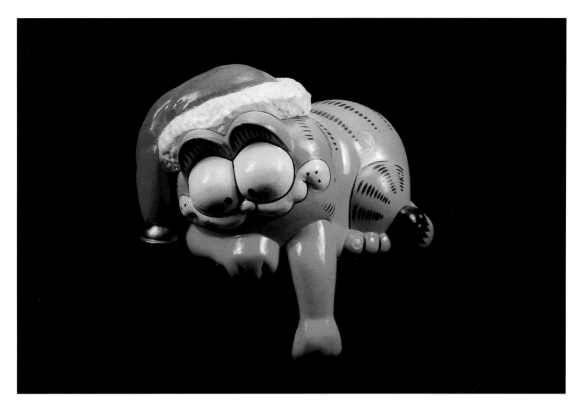

Stocking hanger, plastic. Manufacturer unknown. Garfield is wearing a Santa hat. $50-65.

Bib, cotton/vinyl. Remco Baby, Inc., early 1990s. Caption: "Baby's first Christmas." $5-10.

Bib, cotton/vinyl. Remco Baby, Inc., early 1990s. Caption: "Merry Christmas." $5-10.

Figurine, plastic. Enesco Corporation. Garfield is wearing deer antlers. Titled: "Happy Holly Deers!" 8". $20-35.

Stickers. 3M/Post-it. Removable stickers. Three different versions of Garfield with Christmas themes. Two sheets per package. $5-10 each.

Container, metal. Manufacturer unknown. Top view shown. Garfield is giving a present to Pooky. $15-20.

Candle. Manufacturer unknown. Garfield is hugging a candy cane. $5-10.

Mug, ceramic. Enesco Corporation. Caption: "Christmas tip no. 3...tell 'em your size." $5-10.

Figurine, ceramic. Enesco Corporation. Caption: "Cheers." 4.5." $20-35.

Bibliography

Davis, Jim. *The Garfield Trivia Book*. New York: Ballantine Books, a division of Random House, Inc., 1986.

Karl, Denise. *The Garfield Connection* newsletter. Armonk, New York. Monthly distribution, 1997.

Korbeck, Sharon. *Toy Shop* magazine. Iola, Wisconsin: Krause Publications, September 1997.

Maim, Jim. *Collectible Toys & Values* magazine. Ridgefield, Connecticut: Attic Books, Ltd., September, 1993.

Mittelbach, Margaret. *Country Accents Collectibles, Flea Market Finds* magazine. New York, New York: GCR Publishing Group, Inc., Spring, 1996.

Mittelbach, Margaret. *Country Accents Collectibles, Flea Market Finds* magazine. New York, New York: GCR Publishing Group, Inc., Spring, 1997.

http://www.ara-animation.com
http://www.garfield.com